INCREDIBLE ANIMAL FACE-OFFS

ANIMAL MIGRATOR MATCH-UP

RACHAEL L. THOMAS

Consulting Editor, Diane Craig, M.A./Reading Specialist

Super Sandcastle

An Imprint of Abdo Publishing
abdobooks.com

abdobooks.com

Published by Abdo Publishing, a division of ABDO, PO Box 398166, Minneapolis, Minnesota 55439.

Printed in the United States of America, North Mankato, Minnesota
102019
012020

THIS BOOK CONTAINS RECYCLED MATERIALS

Design: Sarah DeYoung, Mighty Media, Inc.
Production: Mighty Media, Inc.
Editor: Jessica Rusick
Cover Photographs: Shutterstock Images
Interior Photographs: Getty Images/iStockphoto, pp. 11, 12, 13, 17; iStockphoto, pp. 4 (top right), 5 (top left), 20 (top), 21 (top); Shutterstock Images, pp. 4, 5, 6, 7, 8, 9, 10, 13 (right), 14, 15, 16, 18, 19, 20, 21, 22; Yva Momatiuk and John Eastcott/Minden Pictures, p. 23

Library of Congress Control Number: 2019943209

Publisher's Cataloging-in-Publication Data
Names: Thomas, Rachael L., author.
Title: Animal migrator match-up / by Rachael L. Thomas
Description: Minneapolis, Minnesota : Abdo Publishing, 2020 | Series: Incredible animal face-offs
Identifiers: ISBN 9781532191954 (lib. bdg.) | ISBN 9781532178757 (ebook)
Subjects: LCSH: Animal migration--Juvenile literature. | Animals--Migration--Juvenile literature. | Animals, Habits and behavior of--Juvenile literature. | Social behavior in animals--Juvenile literature.
Classification: DDC 591.525--dc23

CONTENTS

BATTLE OF THE MIGRATORS

MONARCH BUTTERFLY

The animal kingdom is full of stars. But some animals stand out. These animals are the best **migrators**.

Animals travel many miles to escape the cold. They also travel to find food.

Animal adventurers are all around us. But what if you matched them up in face-offs? Which animal would be the best migrator?

ARCTIC TERN

WILDEBEEST
SOCKEYE SALMON
RED CRAB
HUMPBACK
WHALE

FLIGHT FIGHT

Arctic terns are outstanding fliers. Monarch butterflies are too! But which animal would win a flight fight?

ARCTIC TERN

FROZEN FLIER

This bird flies between the North and South Poles. Put your hands together for the Arctic tern!

Tail feathers spread wide for smooth gliding

ARCTIC TERN STATS

JOURNEY

Up to 56,000 miles (90,123 km) a year

FOOD

Small fish, **crustaceans**, and insects

SIZE

26 to 30-inch (66 to 75 cm) wingspan

AN ARCTIC TERN'S WINGSPAN IS LONGER THAN AN AVERAGE NEWBORN BABY.

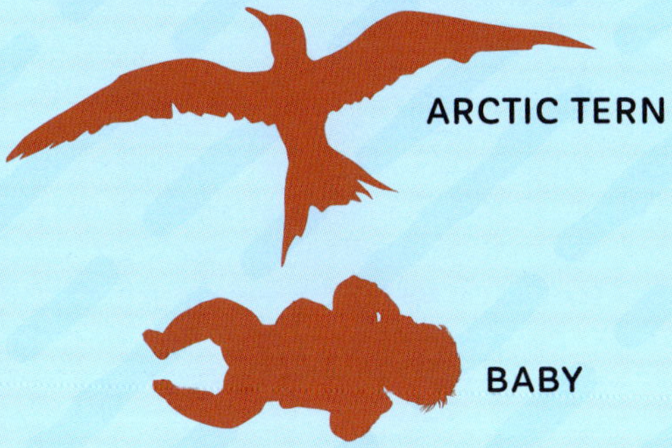

MONARCH BUTTERFLY

FAMILY FLIGHT

Generations of this insect work together to **migrate**. Say hello to the breathtaking monarch butterfly!

MONARCH BUTTERFLY STATS

JOURNEY
3,000 miles (4,828 km)

FOOD
Milkweed plants

SIZE
4-inch (10 cm) wingspan

A MONARCH BUTTERFLY'S WINGSPAN IS SHORTER THAN A DOLLAR BILL.

BUTTERFLY

DOLLAR BILL

FLIGHT FIGHT

ARCTIC TERN VS MONARCH BUTTERFLY

ARCTIC TERN

Arctic terns breed in noisy groups. Before **migration**, the birds fall silent. Then they take off together!

MIGRATION MOTIVATION

Arctic terns nest in the Arctic during summer. Then they journey to Antarctica. They arrive there as a new summer begins!

TOURIST HUBS

Arctic terns migrate over the open sea. They are easiest to view when they nest on land.

MONARCH BUTTERFLY

Millions of monarchs **migrate** each fall. The butterflies lay eggs along the way. New butterflies start where their parents left off.

MIGRATION MOTIVATION

Monarch butterflies fly south to Mexico for winter. Then they fly back north.

TOURIST HUBS

Monarchs land on Mexican oyamel fir trees. They coat these trees by the millions.

WAR OF THE WAVES

Humpback whales are coastal champions. Red crabs are too! But which animal would win a war of the waves?

HUMPBACK WHALE

MIGHTY MIGRATION

This deep-sea diver makes one of the longest **migrations** of any **mammal**. Meet the humpback whale!

HUMPBACK WHALE STATS

JOURNEY

About 3,000 miles (4,828 km) each way

FOOD

Krill and small fish

SIZE

39 to 52 feet (12 to 16 m) long

A HUMPBACK WHALE IS ABOUT AS LONG AS A SCHOOL BUS.

RED CRAB

MOON WATCHER

This **crustacean** follows the moon to **migrate**. Get on your feet for the red crab!

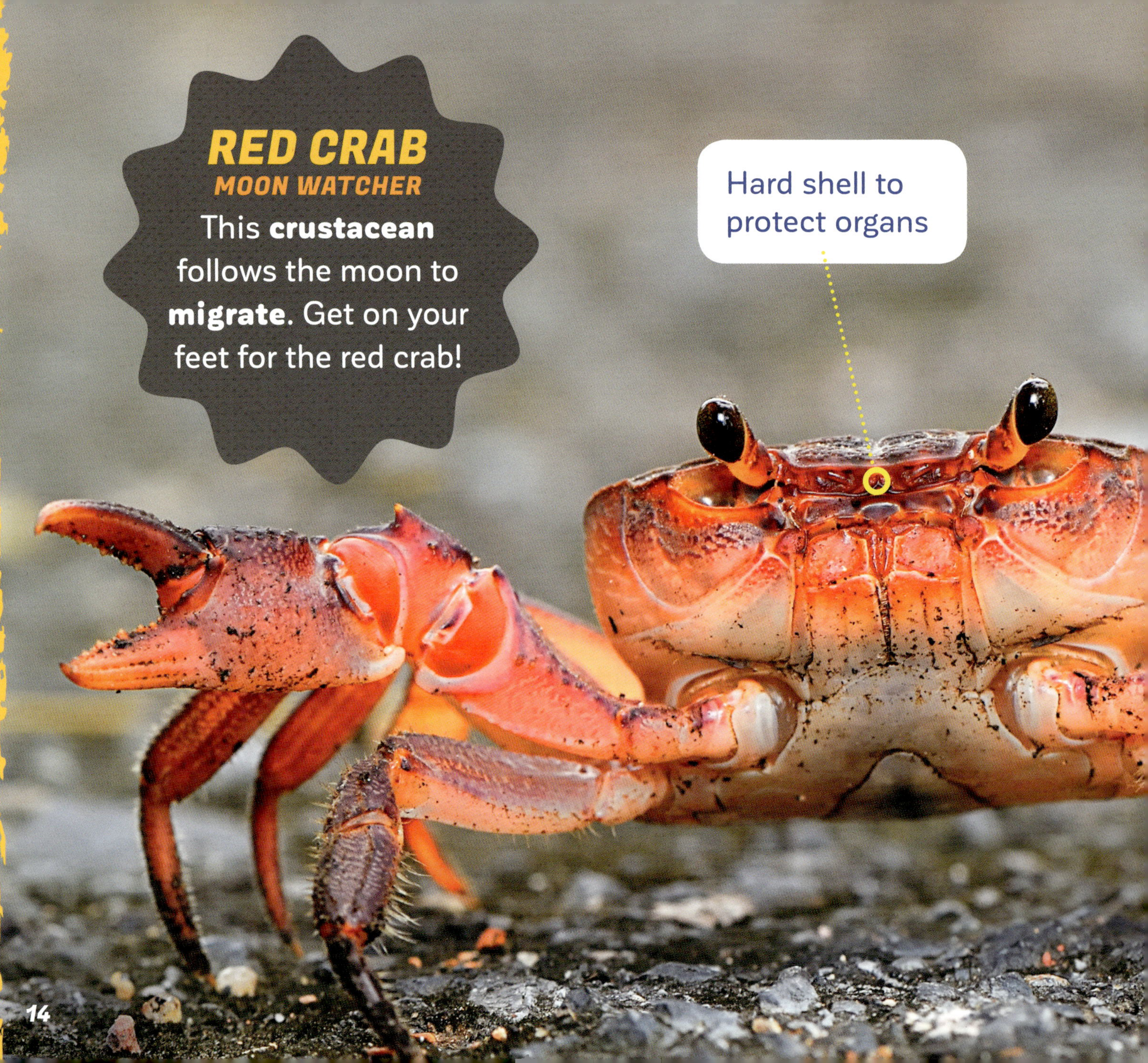

RED CRAB STATS

JOURNEY
About 5 miles (8 km) from forest to shoreline

FOOD
Fallen leaves, seeds, fruit, and flowers

SIZE
About 5 inches (13 cm) wide

A RED CRAB IS LARGER THAN A TEACUP.

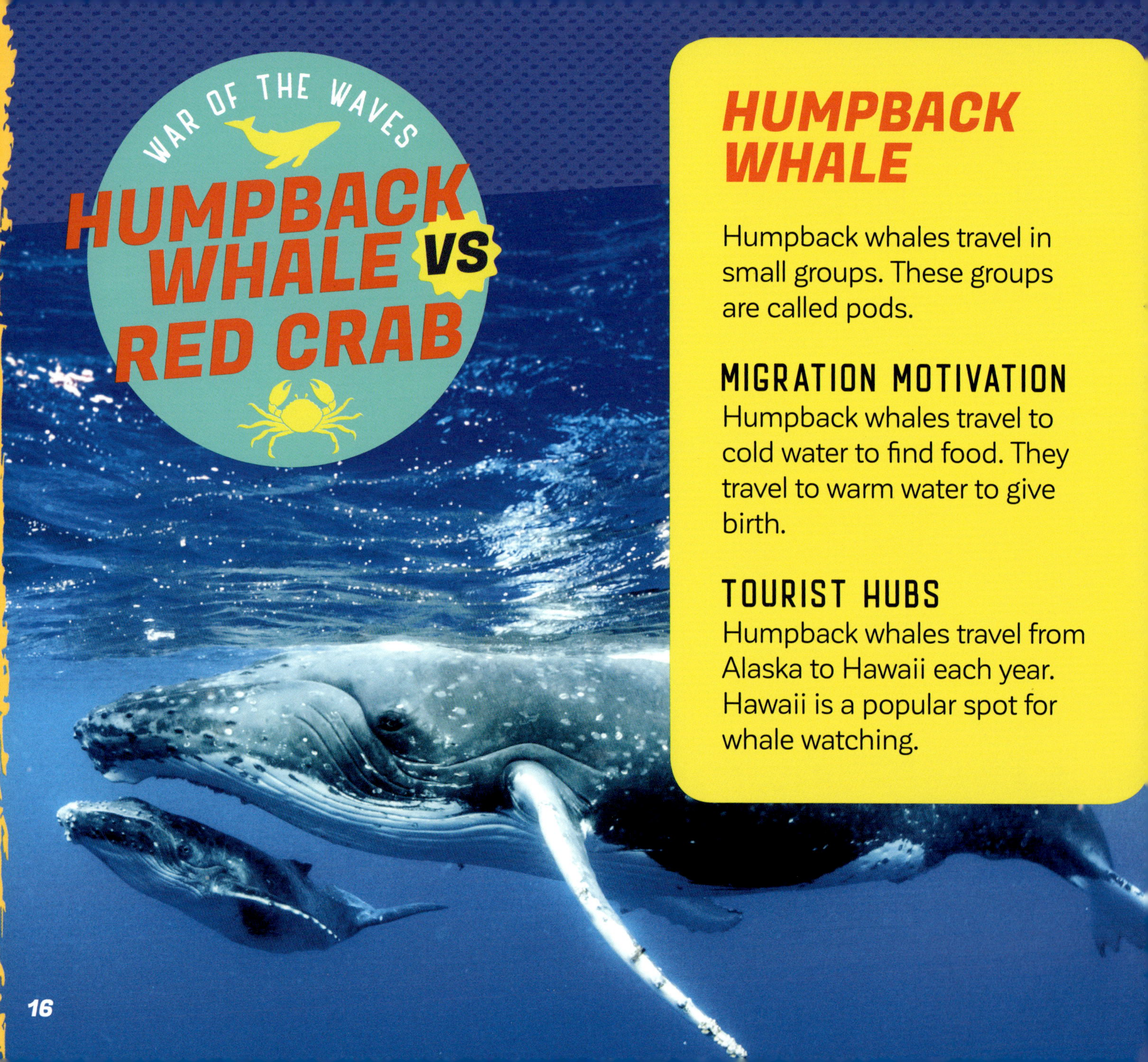

HUMPBACK WHALE

Humpback whales travel in small groups. These groups are called pods.

MIGRATION MOTIVATION

Humpback whales travel to cold water to find food. They travel to warm water to give birth.

TOURIST HUBS

Humpback whales travel from Alaska to Hawaii each year. Hawaii is a popular spot for whale watching.

RED CRAB

Red crabs follow the moon's calendar. The moon affects the ocean's rise and fall. Crabs lay eggs when the water is just right.

MIGRATION MOTIVATION

Red crabs **migrate** to lay eggs. They lay their eggs into the sea. Baby crabs emerge from the waves. Then, they journey back to Christmas Island's forests.

TOURIST HUBS

All red crabs live on Christmas Island! This island is part of Australia.

FRESHWATER FACE-OFF

Wildebeest brave river currents to **migrate**. Sockeye salmon do too! But which animal would win in a freshwater face-off?

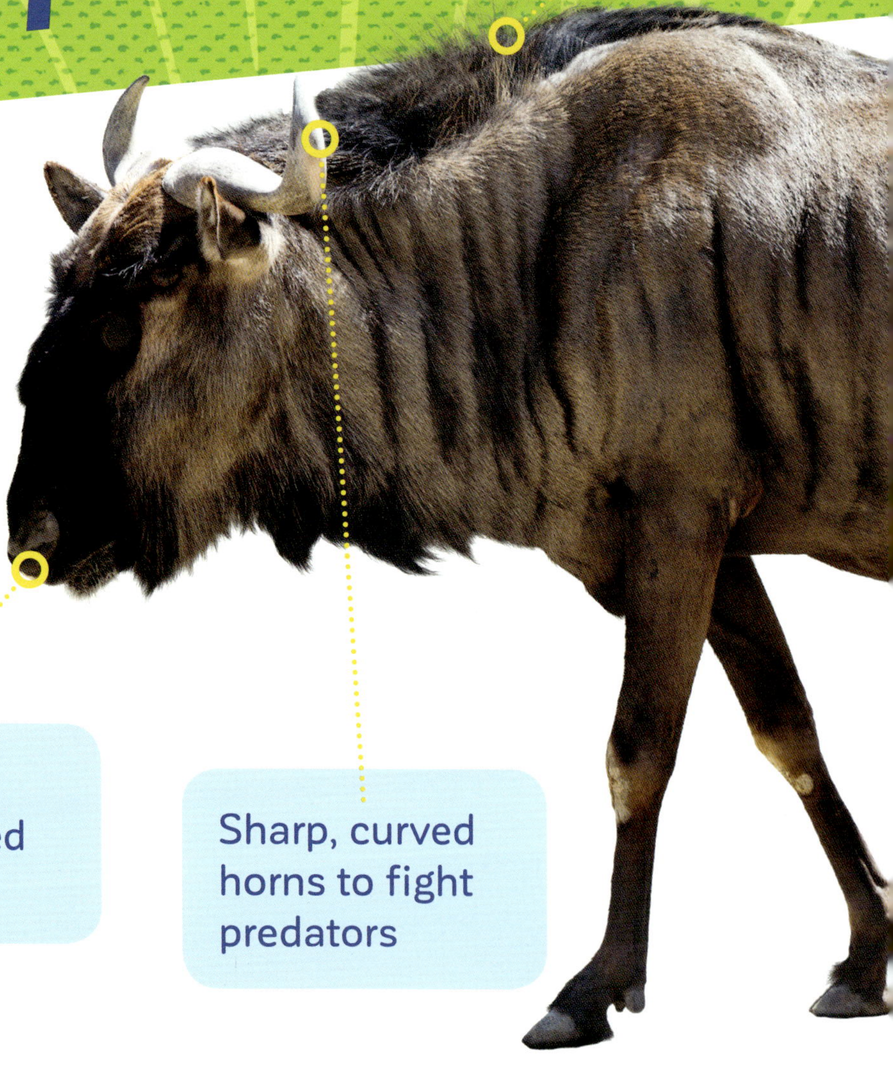

WILDEBEEST

STRONG SWIMMER

This African adventurer searches the **Serengeti** for food and water. Put your hands together for the wildebeest!

WILDEBEEST STATS

JOURNEY

1,000 miles (1,609 km) each year

FOOD

Grass and leaves

SIZE

Up to 4½ feet (1.4 m) tall and 600 pounds (272 kg)

A WILDEBEEST IS SHORTER THAN AN AVERAGE MAN.

SOCKEYE SALMON

SWIMMING SHAPESHIFTER

This fish can find its way home after traveling miles to the open ocean. Say hello to the sockeye salmon!

The sockeye changes when it returns to fresh water to **spawn**.

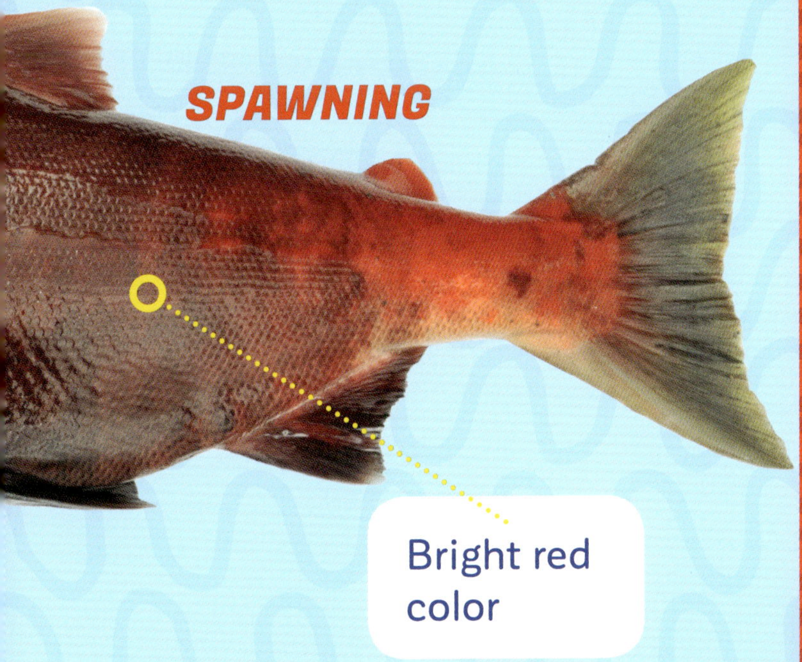

SALMON STATS

JOURNEY
Up to 900 miles (1,448 km)

FOOD
Plankton, insects, small fish, and squid

SIZE
$1\frac{1}{2}$ to $2\frac{1}{2}$ feet (0.5 to 0.8 m) long

A SOCKEYE SALMON IS LONGER THAN A MAILBOX.

FRESHWATER FACE-OFF

WILDEBEEST VS SOCKEYE SALMON

WILDEBEEST

A **migrating** wildebeest herd is large. It can have 2 million animals! A herd is also long. It can stretch up to 25 miles (40 km)!

MIGRATION MOTIVATION

Wildebeest migrate to follow rain. More rain means better grasses to eat.

TOURIST HUBS

Wildebeest cross many rivers during migration. In summer, the herds cross the Mara River in Tanzania. It is full of crocodiles!

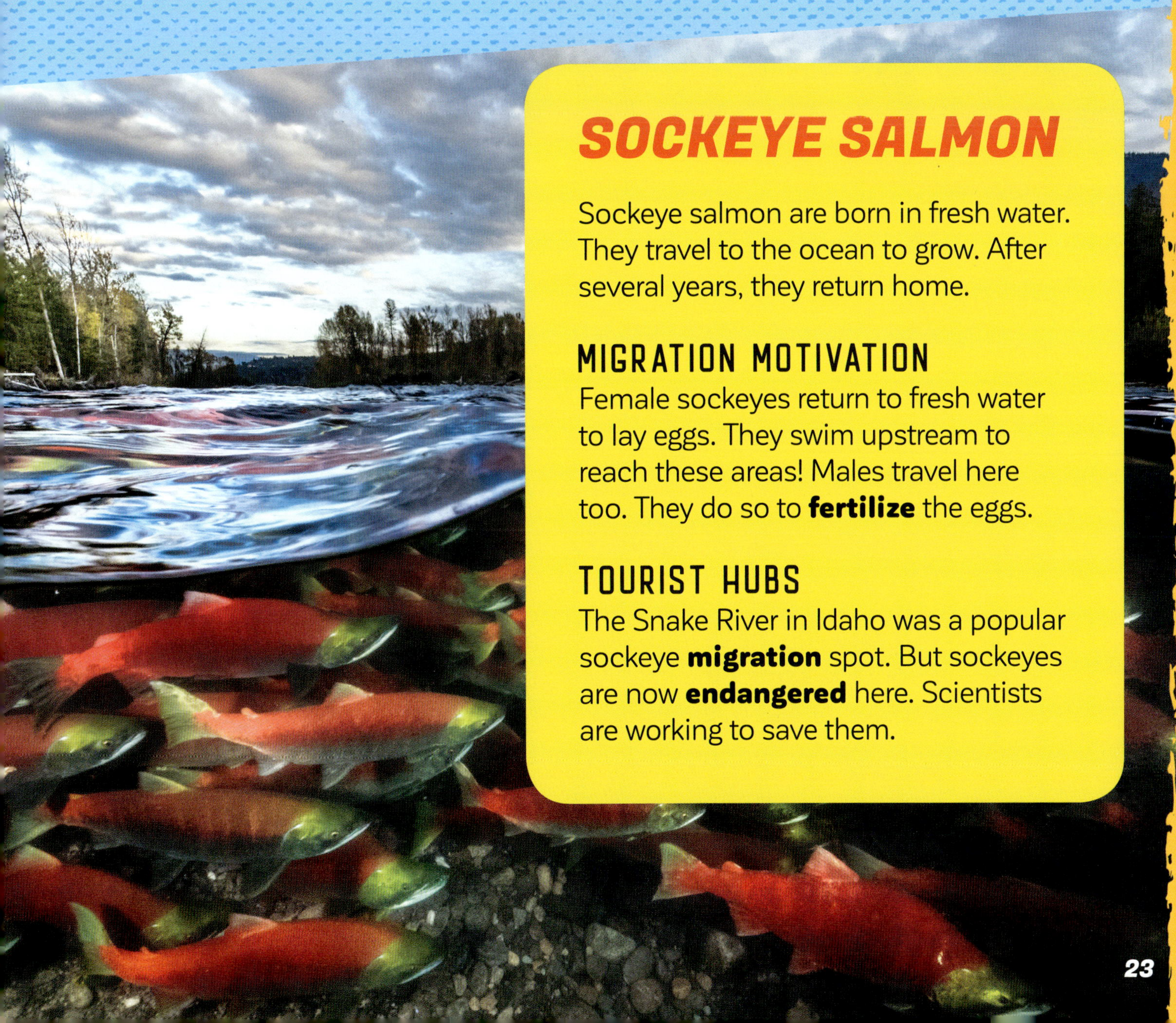

SOCKEYE SALMON

Sockeye salmon are born in fresh water. They travel to the ocean to grow. After several years, they return home.

MIGRATION MOTIVATION

Female sockeyes return to fresh water to lay eggs. They swim upstream to reach these areas! Males travel here too. They do so to **fertilize** the eggs.

TOURIST HUBS

The Snake River in Idaho was a popular sockeye **migration** spot. But sockeyes are now **endangered** here. Scientists are working to save them.

GLOSSARY

crustacean—a sea creature, such as a lobster, crab, or shrimp, that has a hard, external skeleton.

endangered—having few left in the world.

fertilize—to make fertile. Something that is fertile is capable of growing or developing.

krill—very small creatures that live in the ocean.

mammal—a warm-blooded animal that has hair and whose females produce milk to feed their young.

migrate—to move from one area or climate to another, usually at about the same time each year. This movement is called migration. An animal that migrates is a migrator.

nostril—an opening in the nose.

plankton—very tiny organisms that live in water.

Serengeti—an area of woodlands and grasslands in eastern Africa.

spawn—to produce or deposit eggs.